Bibliographic information published by the German National Library:

The German National Library lists this publication in the National Bibliography; detailed bibliographic data are available on the Internet at http://dnb.dnb.de .

Imprint:

Copyright © 2018 GRIN Verlag
Print and binding: Books on Demand GmbH, Norderstedt Germany
ISBN: 9783668801868

This book at GRIN:

https://www.grin.com/document/440897

Philip Mensah

Incorporation of Solid Waste Management Skills in the Education Curriculum of India for a sustainable Environment and Economic Effectiveness

GRIN Verlag

Inhalt

Abstract

Education of the people globally is expected to help solve societal and environmental problems, improve, and manage way of life. However, solid waste management has become a challenge in many countries not excluding India despite intensive education. Solid waste in India has been considered as low priority and is causing unsustainable environment (pollution of air and water bodies, ill health, flood, economic instability etc.). A well designed education curriculum for the people can help best manage solid waste even though it is unavoidable. This study seek to incorporate the lapses of solid waste management skills: handling, collecting, disposing, storage, transporting, recycling, dumping etc. in the Secondary and senior Secondary education curriculum of India as compulsory and suggest to the stakeholders' most appropriate solid waste management skills to be provided by the Secondary and senior Secondary curriculum of India. At the Secondary level curriculum, these skills are missing from Social Science syllabus and not well detail in the Environmental Science of the Senior Secondary level curriculum. In all the population selected for the eight towns, Forty (40) interviews will be conducted in each of the towns and eight hundred (800) questionnaires will be issued out using random sampling technique. One hundred (100) questionnaires will be administered to the respondents in each of the towns: Goraya and Philour of Jalandhar District, Khaira and Samrala of Ludhiana District, Phagwara and Sultanpru of Kapurthala, and Shahid Bhagat Singh (S.B.S), all in the Punjab State; students, teachers, workers of waste management organizations, and education experts.

Key words: Solid waste management skills, Secondary and Senior Secondary Education Curriculum, Sustainable environment, and Economic effectiveness.

INTRODUCTION

Background of the study

Much is happening internationally to promote environmental education (Donkor, 1996, p.1). The scares resources, vast increasing population, rapid urbanization, and increasing global industrialization have called for protection of the environment through education (Hazra & Goel, 2008). Environmental education can be defined as "process of providing learning experiences to obtain knowledge, understanding, skills, and awareness with desirable attitudinal changes about man's relationship with his natural and man-made surroundings" (Lakshmi et.al, 2006, p. 10). Protection of public health and environment from bad condition must be responsibility of everyone because it enhances reduction or elimination of adverse negative impacts on the environment and human health, supports economic development and improved quality of life. In the Asian and African continents, several practices are making the environments out of shape for living even though efforts to manage it has been made by some experts (Zhu et.al 2007). Earlier 1990s, were no laws in the books of many Asian countries about waste generation, disposal, and treatment (Carolina & Dewitt, 2003). Nevertheless, some national governments: India, Pakistan, Bangladesh, etc. Have started regulating management and disposal of solid waste (Ray, 2008).

Solid waste management is a major challenge in every human society because daily activities of human beings bring solid waste and pose risks to the environment and the public health due to the way these wastes are handled, stored, collected, and disposed of (Zurbrugg, 2002). Effective solid waste management is very beneficial to allow healthy living conditions for the entire population. The fact to maintain healthy environmental conditions for the people has been accepted by a lot of governments but many municipalities find it so hard to provide services for healthy living. An India activist, Mahatma Gandhi famously said "Sanitation is more important than independence". Recently, solid waste has been noted as one of the main challenges causing flood, health problems and draining the economy in India (Ray, 2008). Over 90% of the solid waste produced in India is directly disposed on land in unacceptable mode (Das et al., 1998).

Management of solid waste in India has always been considered as low priority area, (Kansal, 2002). All effort to ensure effective management of solid waste in India is being put in place by experts, Patriotic citizens, international organizations' particularly the Government of India (GOI) through provisions of grants and loans to state governments under the fourth 5-Year plan of 1969-74 (Talyan, et.al, 2008). Volume of solid waste is moving up "as economic growth drives more and more people from the rural hinterland to the urban areas in India, spawning newer consumption patterns and social linkages. The large metropolises of India now generate more than 6,000 tons of solid waste per day, and Delhi alone generates more than 3,500 tons. By the year 2030, India will probably generate more than 125,000 metric tons of waste every year" (Gupta, 2004 cited by Ray, 2008)

In the research of Marcin, et al (1994), the ideal way to help manage the solid waste is source reduction thus promotion of market mechanisms to effect source reduction through fees and tax

Incentives, ensuring of mandatory standards and regulation, and the use education and voluntary compliance with policies by business and consumers.

Considering the education sector to help overcome the challenge is the use of environmental education formal or informal. If the problem of solid waste in India is not falling as a result of increasing number of the people in cities then it gives alarm that current curriculum of environmental education is not being helpful to the management of the menace. A study shows increases of 49% population and 67% for Solid Waste same time in India, (UNEP, 2001). The over 1.21 billion population of India representing 17.5% global population according to Census of India, 2011 is contributing to the high rate of urbanization as well as improper planning and poor financial condition has made solid waste control in Indian cities a difficult task (Kaushal et.al, 2012). The Cities population stated by the Census were: Greater Mumbai (18.4 million), Delhi (16.3 million), Kolkata (14.1 million), Chennai (8.7 million), and Bangalore (8.2 million) (Kaushal et.al, 2012). The assessment of Kumar et.al, (2009), provides the population of Indian cities against waste generation rate (Kg) per capital per day as:

S/N	Population	Waste Generation Rate Kg/Capital/day
1.	Cities with a population < 0.1million	0.17 - 0.54
2.	Cities with a population 0.1- 0.5million	0.22 – 0.59
3.	Cities with a population 1- 2 million	0.19 – 0.53
4.	Cities with a population > 2 million	0.22 – 062

There are so many factors that lead to the generation of waste. According to Sujauddin et al. (2008) the generation of waste is influenced by family size, education level and the monthly income. Ekere et al., (2009) express gender, peer influence, land size, location of household and membership of environmental organization explain household waste utilization, and separation behavior have influence on waste generation.

Sources and Types of Solid Waste in India Source

Sources	Typical waste generators	Components of solid waste
Residential	Stores, hotels, restaurants, markets, office buildings	Food wastes, paper, cardboard, plastics, textiles, glass, metals, ashes, special wastes (bulky items, consumer electronics, batteries, oil, tires) and household hazardous wastes
Commercial	Stores, hotels, restaurants, markets, office buildings	Paper, cardboard, plastics, wood, food wastes, glass, metals, special wastes, hazardous wastes
Institutional	Schools, government center, hospitals, prisons	Paper, cardboard, plastics, wood, food wastes, glass, metals, special wastes, hazardous wastes
Municipal services	Street cleaning, landscaping, parks, beaches, recreational areas	Street sweepings, landscape and tree trimmings, general wastes from parks, beaches, and other recreational areas

Source: Alonso & Themelis (2011)

Larijani & Yeshodhara (2008), ague that, awareness, skills, Values, Knowledge, and attitudes learnt through education must assist people to lead a desired quality of life. Currently, environmental education in India through the Central Board of Secondary Education (CBSE) provides idea about the environment but not enough as far as solid waste management issues are concern. It is expected to provide solid waste management skills such as handling, collecting, disposing, storage, transporting, recycling, dumping etc.

The above revelations of the improper management of solid waste in India have called for the need to manage the problem by inclusion of solid waste management skills in the Secondary and senior Secondary education curriculum by equipping each subject teacher at the level to impart the skills in his classroom teachings.

Research Gap

20th and early 21st century events indicate that waste in whatever form or classification- solid, liquid or toxic have become a major consequence of modernization and economic development (Tsiboe & Marbell, 2004) According to Pervez & Kafeel (2013) effective solid waste management have to be undertaken to ensure that it does not affect the environment and not cause health hazards to the people living in a community.

It is quite unfortunate that research works on solid waste management have concentrated on public awareness, the use of technologies, types of solid waste, and other possible eliminating factors and not provision of managerial skills through classroom education. Everyone needs managerial skills and interest in waste related occupations to help manage the menace. Kironde (1999) pointed out that, the recent upsurge in waste disposal problems stems from the fact that "attitudes and perceptions towards wastes and the rating of waste disposal issues in people's minds and in the scheme of official development plans have not been adequately considered". India's Secondary and Senior Secondary education curriculum on Environmental Science is really doing its best however, meeting it goal regarding solid waste management is a problem due to:

1. Its omission in the Social Science syllabus at the Secondary education level is provided.
2. Its Optional and unspecific in the Environmental Science at Senior Secondary School level on the matters relating to solid waste managerial skills for effective management of the menace.
3. Its inability to provide solid waste management skills to every classroom teacher, no matter his specialization through a seminar to enable the teacher do impartation during classroom teaching.

Basically, this study seeks the best way to manage solid waste through incorporation of solid waste management skills in the CBSE's Social Science subject at the Secondary education curriculum and suggesting solid waste management skills in Environmental Science as compulsory subject at the Senior Secondary education curriculum of India for a sustainable environment and economic effectiveness. Here, every teacher regardless of his or her subject area is expected also to go through training to become knowledgeable (skillful) in solid waste management to gradually impart these skills into learners during teaching. This does not mean that solid waste management skills should be a separate subject at the various levels of Secondary and Senior Secondary education. "We cannot always build the future for our youth, but we can build our youth for the future," former US President, Franklin D. Roosevelt. Sizeable number of the youth are into education and are pillars of this country, and must be prepared from this level for skills and interest in solid waste management to help manage the problem.

According to the former Franklin D. Roosevelt again, 'a nation's future depends on its youth since they (youth) will be the future decision makers and stakeholders of the developmental process. One of the daunting tasks is to devise ways to enable our future generations to cope with the perils of environmental catastrophe'. Acquisition of solid waste management skills through Secondary and Senior Secondary education must be right tool for managing the menace of solid waste (Desa, et. al, 2012). The Kothari commission (1964-66) proposed that "basic education had to offer Environmental Education and relate it to the life needs and aspirations of the people and the nation". The report recommended that " the aims of teaching science in the Secondary schools should be to develop proper understanding of the main facts, concepts, principles and processes in physical and biological environment" Environmental education at the Secondary and Senior Secondary levels according the report was treated in a different way. Environmental education is an essential part of every student's learning. It helps to encourage awareness of the environment, leading to informed concern for active participation in resolving environmental problems".

Statement of the Problem

The sight and smell of inadequately managed solid waste in India create a major discomfort to residents and visitors. Pollution of water resources increases the technical difficulty and cost of providing water supplies and the environmental health situation also has serious health impact, with attendant social and economic costs. Flooding with its associated damage to public infrastructure and private property increases with improper solid waste management.

Economic activities, large increase in urbanization and population in India have made the impact of the society's solid waste very noticeable (Agarwal et.al, 2005). Gupta et.al (1998), View the collection, transportation and disposal of solid waste in India as unscientific and chaotic leading to overflowing landfills, serious environmental implications in terms of ground water pollution, and contribution to global warming. About 80% of the total budget of all municipal corporations is accounted for by the salaries of sanitation workers engaged in road sweeping and related activities in India (Gupta et.al 1998),

In view of this, the statement of the problem of this study: incorporation or integration of solid waste management skills in the Secondary and Senior Secondary education curriculum of India has led to less management of solid waste in India and could effectively be resolved with instilled solid waste management skills in the Indian youth. The uncared environment according to Hazra & Goel (2008), is posing the problem of parasites, cholera, diarrhea and malaria's prevalence of parts of India have been as a result of unsanitary conditions in and around us. "Common infectious diseases like malaria, intestinal worms, and upper respiratory infections are among the most common health problems reported at the out-patient facilities in most states in India, and majority of these cases are residents in and around the slums".

Purpose of the study

The study purposely assesses current solid waste management challenges on the environment of India, evaluates efforts being made by Secondary and Senior Secondary education curriculum in the area of solid waste management skills, provides incorporation of solid waste management skills in Secondary and Senior Secondary curriculum on the environment for sustainable environment and economic effectiveness of India, and techniques for incorporating the skills of solid waste management in Secondary and Senior Secondary curriculum, and finally gives suggestions and recommendations on waste management on the environment.

Research Objectives

The main objective of the study is to examine how improved Secondary and Senior Secondary education Curriculum of the Environmental Science subject under CBSE of India can enhance the solid waste management. The specific objectives of the study are:

1. To explain the concept incorporation of solid waste management skills in Secondary and Senior Secondary curriculum for sustainable environment and economic effectiveness of India.
2. To evaluate the current solid waste management practices of India.
3. To assess the impact of incorporation of solid waste management skills in Secondary and Senior Secondary curriculum of India.
4. To provide means of acquiring solid waste management skills in India through education.
5. To provide suggestions and recommendations on solid waste management in India.

Research Questions

1. What are the current solid waste management practices in India?
2. What role is education currently playing in solid waste management of India?
3. What impact can incorporation of solid waste management skills in Secondary and Senior Secondary curriculum have on the environment India?
4. How can one acquire better solid waste management skills in India through education?

Significance of the study.

A lot of significance can be attributed to this research. Few are provided as follows; firstly, it would help individuals to learn the basic concepts of incorporation of solid waste management skills in Secondary and Senior Secondary education for sustainable environment and economic effectiveness of India, effects, and benefits of controlling waste in India. Secondly, it will help individuals, organizations, and India government to be abreast with the contribution of Secondary and Senior Secondary education in controlling environmental waste. Thirdly, Education at this level will receive much attention and investment to assist in managing the solid waste problem of India, and create awareness of the current practices of controlling waste for sustainable environment and economic effectiveness in India. Fourthly, credible and safe approaches to the disposing or managing, collecting, transporting, processing, monitoring, recycling of waste materials and many more will be established. Sixthly, it will educate people on the protection of public health and environmental waste control. Resource management through waste prevention, recycling and reuse practices, material recovery etc. thus to maintain good health. Seventhly, Secondary and Senior Secondary education curriculum experts will identify the need to amend Environmental Science education in the current curriculum of CBSE. Lastly, it will help individuals and organizations realize the urgent need of sponsoring Education in India.

Scope of the study

Geographically, the study area will cover Goraya and Philour of Jalandhar District, Khaira and Samrala of Ludhiana District, Phagwara and Sultanpru of Kapurthala, and Shahid Bhagat Singh (S.B.S), all in the Punjab State. These cities have been chosen because they are familiar to the researcher and very good for generalization. Contextually, the study will focus on Environmental Science education curriculum of Secondary and Senior Secondary levels under the CBSE and domestic solid waste management.

Limitation of the study

Despite the fact that this study will aim at the incorporation of solid waste management skills in the Secondary and Senior Secondary curriculum of India for a sustainable environment and economic effectiveness, its data collection will not cover the entire Country. Again, the study is limited to Environmental Science education curriculum of Secondary and Senior Secondary levels under the CBSE, domestic solid waste management, and selected students, teachers, and environmental workers in the identified towns of India due to insufficient funds, materials, and time resource.

Organisation of the study

This study will comprise five chapters. These are outlined as follows: Chapter one is the introduction. It deals with background of the study, the statement of the problem, the purpose of the study, research objectives, significance of the study, research questions, and, limitation of the study, delimitation of the study as well as the organization of the study.

Chapter two will focus on the review of the related literature. It will generally looks at the incorporation of solid waste management skills in Secondary and Senior Secondary education for sustainable environment and economic effectiveness of India.

Methodology of the studies will be in chapter three. This chapter will describes; the research design, population and sampling procedures, source of data collection, instrumentation, pre-test of the study, data analysis procedure, and ethical considerations. Chapter four will provide the result of the study and describes the data presentation and how they were analyzed in relation to the problem. Chapter five will comprises summary of findings, conclusion, recommendation(s) and suggestions for further research.

REVIEW OF LITERATURE

This chapter represents a review of literature on incorporation of solid waste management skills in the Secondary and Senior Secondary education curriculum of India for a sustainable environment and economic effectiveness of the following areas: The concept of incorporation solid waste management skills in the Secondary and Senior Secondary curriculum of India for a sustainable environment and economic effectiveness, Current solid waste management practices of India, Current impact of the Secondary and Senior Secondary education curriculum on solid waste management skills of India, and Practical ways to manage solid waste.

The concept: Incorporation of solid waste management skills in the Secondary and Senior Secondary education curriculum of India for a sustainable environment and Economic effectiveness

Incorporation is to include something as part of a group, system, plan vice versa (Mayor 2009). Similar definition was given by Dictionary, C. (2008), as to include something as part of a larger, which is also not different from the meaning given by Hornby et.al (1995) as to include something so that it forms a part of something. The main synonym of the word incorporation in this study is integration which can also be defined as: the process of bringing together two or more things into one (Jones, 2011). According to Singh Y. K. (2010), integration is a curriculum to be followed by using a single theme to teach a variety of subjects. Solid is a firm and stable in shape, not liquid or fluid (businessdictioary.com). According to Dictionary, C. (2008), Solid is hard or firm keeping a clear shape. Waste is more easily recognized than defined. Something can become waste when it is no longer useful to the owner or it is used and fails to fulfill its purpose (Gourlay, 1992 cited by Freduah, 2004). There are basically two types of waste namely liquid and solid waste. But for the purpose of the study, the focus is on solid waste. This is discussed in the next section of the review.

Solid Waste

Solid waste are all forms of none liquid or stable in shape items which are no longer useful to their owners and these include the types: Biomedical waste or hospital waste which is very infectious waste. Municipal solid waste consists of household waste, construction and demolition debris, sanitation residue, and waste from streets. This garbage is generated mainly from residential and commercial complexes. Leblang (2017) explained Solid waste as refers to the range of garbage arising from animal and human activities that are discarded as unwanted and useless. Solid waste is generated from industrial, residential and commercial activities in a given area, and may be handled in a variety of ways. Solid waste is any material that arises from human and animal activities that are normally discarded as useless or unwanted (Tchobanoglous et al 1993). According to Zerbock (2003), solid waste includes non-hazardous industrial, commercial and domestic waste including: • household organic trash, street sweepings, institutional garbage, and construction wastes. The Ghana Innovation Market Place (2009) popularly known as 'GIM' defines also see solid waste as neither wastewater discharges nor atmospheric emissions, arising from domestic, commercial, industrial, and institutional activities in an urban area. Solid waste is the generation of undesirable substances which is left after they are used once Jatput et.al (2009). According to Harold (1957), Management "is an art of getting things done through and with the people in formally organized groups. It is an art of creating an environment in which people can perform and individuals and can co-operate towards attainment of group goals". And to Taylor (1911) "Management is an art of knowing what to do, when to do and see that it is done in the best and cheapest way". Skills is ability or proficiency in execution or performance (international Webster's student dictionary 2009).

Solid Waste Management Skills are competences systematically used to control of generation, collection, storage, transport, separation, processing, treatment, recovery, and disposal of solid waste (BusinessDictionary.com Retrieved March 12, 2018).

Solid Waste Management Skills are disciplines associated with control of generation, storage, collection, transport or transfer, processing and disposal of solid waste materials in a way that best addresses the range of public health, conservation, economics, aesthetic, engineering and other environmental considerations. In its scope, solid waste management includes planning, administrative, financial, engineering and legal functions. Solutions might include complex inter-disciplinary relations among fields such as public health, city and regional planning, political science, geography, sociology, economics, communication and conservation, demography, engineering and material sciences (Leblang 2017).

Secondary and Senior Secondary Education

According to Roy, (2014), the secondary education is of three years duration from age between10-11 years and end at13-14 years. Education at the Secondary level are of basic programs as teaching is more subject-focused. The Senior Secondary education level also use two years for students from age 14 to 16 years. Requirements for admission is completion of upper primary school education. Instruction is more organized along specific subjects.

Curriculum is a term which referred to the lessons and academic the sum total of all the learnings in a school or in a specific course or program. Curriculum is used for courses students learn in the school Hornby et.al (1995). According to the Glossary of Education Reform (2013), Curriculum as "the knowledge and skills students are expected to learn, which includes the learning standards or learning objectives they are expected to meet; the units and lessons that teachers teach; the assignments and projects given to students; the books, materials, videos, presentations, and readings used in a course; and the tests, assessments, and other methods used to evaluate student learning".

Secondary and Senior Secondary Education Curriculum is the sum total of the learnings or specific programs in the Secondary and Senior Secondary education.

India is a republic country in south Asia as the seventh-largest country by area (3287263Km) in the world. According to Wikipedia, India is now the tenth industrialized nation has a population of over 1.2 billion people and shares boundary with Indian Ocean on the south, the Arabian Sea on the southwest, and the Bay of Bengal on the southeast. India is a federal republic governed under a parliamentary system and consists of 29 states and 7 union territories. It is found at latitudes 8° 4% and 37° 6% north, longitudes 68° 7% and 97° 25% east and measures about 3214 km from north to south between the extreme latitudes and about 2933 km from east to west between the extreme longitudes. The country had independence on 15th August, 1947 and using Rupee as currency. The capital town of India is Delhi. "The quantum of waste generated in the country is increasing day-by-day on account of its increasing population and increased GDP, though the civic services have not been expanding proportionately and hence are under tremendous pressure" (Gupta et.al, 1998). Indians major cities are shown on the map that follows.

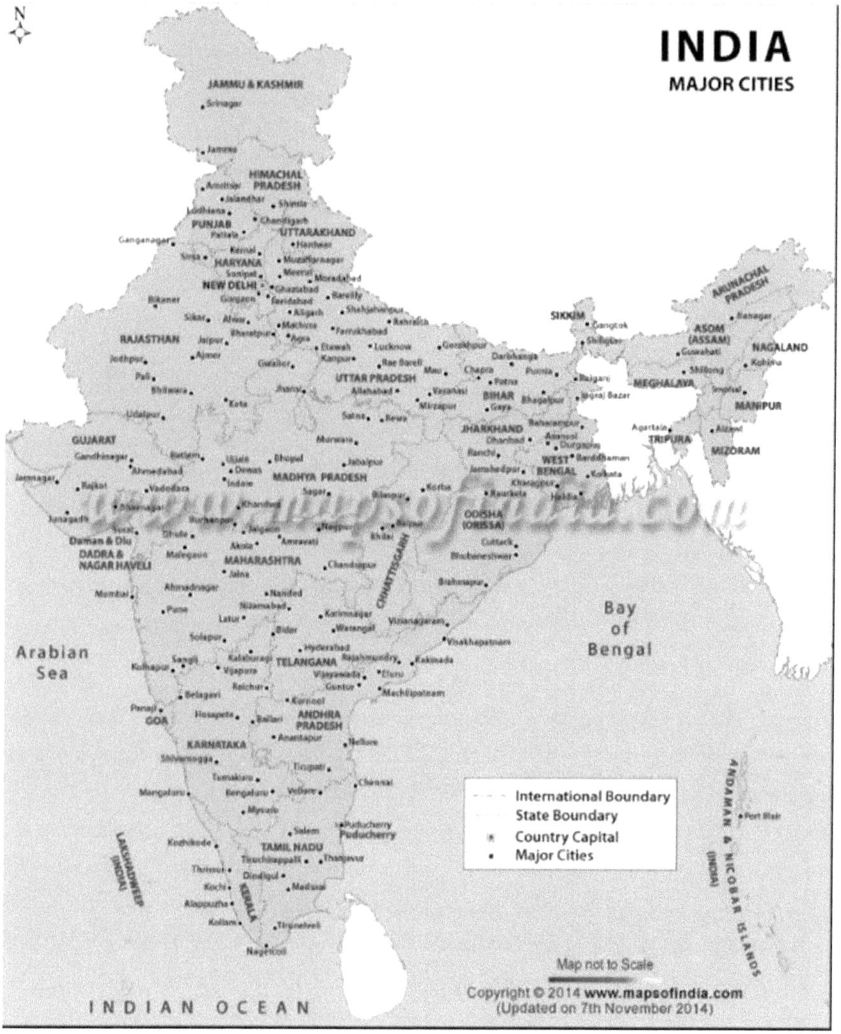

Source https://www.mapsofindia.com/maps/cities/india-map-with-cities.html Retrieved on 25-08-2018

Sustainable is the constant interaction of something. Environment is a term commonly referred to that which is comprised all living and non-living things that occur naturally on earth or some region thereof (Patil & Biradar, 2009). Sustainable Environment is the constant interaction of the environment and mankind without separation (Levins & Lewontin, 1994). Economic Effectiveness is the "measuring the efficacy and purposefulness of a given business activity given by the comparison (ratio) of the value of the obtained effects to the factors invested in order to achieve them" (The General Household Encyclopaedia, (1995). Therefore, the topic: Incorporation of solid waste management skills in the Secondary and Senior Secondary Education curriculum of India for a sustainable environment and economic effectiveness is basically addition of solid waste control proficiencies in the CBSE Social Science syllabus provided at the Secondary level and making of Environmental Science a compulsory syllabus at the Senior Secondary Level of India to ensure good environment with less expenditure.

Current solid waste management practices of India.

Waste management practices so far in India have not suitably included other solutions as collection, treatment, and supply for re-use, reprocessing and final disposal (Zurbrugg, 2002). Solid waste management in the municipalities of India is the duty of the local government (Gupta et.al, 1998). Year 2000 MoEF rules on Solid waste management and recent once issued to ensure proper waste management in India have recently been published (Ministry of Environment and Forests, 2015). MoEF. Authorities in charge are responsible to do the implementation (Kumar et.al, 2017).

The practices have also not delivered the optimum economic and environmental result for now and have not provided enough room to adapt to future pressures (increases in waste quantities and composition). The study of Huysman, (1994), on the topic: waste picking as a survival strategy for women in Indian cities disclose "picking" as another current practice of managing solid waste in India. According to the researcher, there are 25,000 castes street women waste pickers in Bangalore, a south Indian city who work as waste pickers to earn salaries. In the study of Narayana (2009), India is currently using landfilling for waste management. In the 1980's, India established an incinerator plant in Delhi to use it to manage solid waste through the generation of power for the local grid but the operational experience did not become satisfactory as a result of waste content and the high costs of setting up and running the plants. Gupta et.al (1998) ague that India was predominantly using re-use of waste method through selling of waste books, newspapers, magazines, bottles of glass and plastics, metals, cans etc. to manage waste until recently shifted to recycle.

Current impact of Secondary and Senior Secondary Education on solid waste management of India

Specific literature regarding the impact of Secondary and Senior Secondary Education on solid waste management in India have yet not be found. However, Gopal & Anand, (2005) reviews that current impact of Secondary and Senior Secondary Education curriculum on the environment of India as:

1. "Increases public awareness about environmental issues, explore possible solutions, and to lay the foundations for a fully informed and active participation of individual in the protection of environment and the prudent and rational use of natural resources.

2. Provides opportunities to acquire the knowledge, values, attitudes, commitment and skills needed to protect and improve the environment.

3. Encourages pupils to examine and interpret the environment from a variety of perspectives-physical, geographical, biological, sociological, economic, political, technological, historical, esthetic and ethical.

4. Arouses pupil's awareness and curiosity about the environment and encourage active participation in resolving environmental problems, and

5. Environmental education is closely linked to the other cross circular themes of other subject areas".

Practical ways to manage solid waste

According to Kansal (2002), solid waste by setting target for waste reduction, regulatory reforms, and technological interventions. Asmawati et.al (2012) have raised the point that, several universities have successfully implemented a 'greening' university campus; whereby solid waste management programs were carefully planned based on key focus and waste characterizations. It is expected that solid waste management activities in university campus involve the students as part of their learning process. The particular skills and knowledge gained from environmental education would help in changing human behavior towards the environment. Students with some knowledge and skills on environmental education are more motivated to take part in environmental protection activities and plans thus would generate new ideas for the solution of environmental problems. Sharing new information from their activities with families, other adults, and community probably will have some positive implications on solid waste management practices. Monney (2014, May 27), made a point that Metropolitans, Municipals, and District Assemblies (MMDA's) should collaborate with research institutions to undertake studies on solid waste management in their respective areas and furnish them with the outcomes to aid decision making and action planning.

The idea given by Owusu-Sekyere, et.al (2013) on practical ways to manage solid waste is to be regarded it as a great economic resource. Segregation, re-use and recycling of waste at the household levels or point of generation should be supported and funded at the government level. This should be taken as a national assignment with private sector involvement.

PROPOSED RESEARCH METHODOLOGY

Research Methodology is a body of knowledge that enables researchers to explain and analyze methods indicating their limitations and resources, identifying their presuppositions and consequences and relating their potentialities to research advances (Miller 1983).

Field assessment, data collection and analysis for this study will focused on students, organizations, individuals, and environmental experts in India. Test instruments of questionnaire, and observation guide will be prepared and administered at the selected responses. The data collected will be analyzed by using regression, frequency and correlation to find the significant association between various variables before discussion. Recommendations will aim at practical ways of managing solid waste in India with pre-tertiary education.

Research design

A research design is the controlling plan for a research study in which the methods and procedures for the collecting and analyzing information to be collected is specified (www.buseco.monash.edu.au). A research design therefore provides the glue that holds the research project together. It is used to structure the research and show all its major parts work together to address the central research objectives.

Population

A population can be defined as the complete set of subjects that can be studied: people, objects, animals, plants, organizations from which a sample may be obtained (Shao, 1999). It includes all people or items with the characteristic one wish to understand (Wikipedia). Frankfort-Nachmias & Nachmias, (1996), also gave the definition as the entire group or set of cases that a researcher is interested in generalizing. For the purpose and reliability of this study, the population will consists of students, teachers, workers of waste management organizations, and education experts.

Sample and Sampling method

Due to the heterogeneous nature of the samples, a simple random and stratified random sampling technique will be selected for the sample. In all the population selected for the eight towns, Forty (40) interviews will be conducted in each of the towns and eight hundred (800) questionnaires will be issued out using random sampling technique. One hundred (100) questionnaires will be administered to the respondents in each of the towns: Goraya and Philour of Jalandhar District, Khaira and Samrala of Ludhiana District, Phagwara and Sultanpru of Kapurthala, and Shahid Bhagat Singh (S.B.S), all in the Punjab State; students, teachers, workers of waste management organizations, and education experts.

The respondents will comprise of male and female. The questionnaire method will be used and will consist of thirty items (30) formulated. All research questions used a five point Likert scale such as Strongly Disagreed (SD), Disagreed (D), Neither (N), Agree (A), and Strongly Agreed (SA) will be scored as follows: SD=1, D=2, N=3, A=4, SA= 5. The questionnaire items will be divided into five sections: section A consists of personal data such as sex, age, marital status, qualifications and occupations while section B to E formed the body of the questions in the order of the framed hypotheses. The expected least age should be 15 and above with corresponding frequencies.

Proposed tools for data collection

Data collection will be done via administration of interviews and questionnaires in three towns: Goraya and Philour of Jalandhar District, Khaira and Samrala of Ludhiana District, Phagwara and Sultanpru of Kapurthala, and Shahid Bhagat Singh (S.B.S), all in the Punjab State. Students, teachers, workers of waste management organizations, and education experts. The questionnaires will be collected from the respondents upon completion for analysis.

Data preparation

The coding will be adopted. This will involve the use of simple percentage method to code for data. The total responses for each item will be summed. A total score will be assigned accordingly for computation. These scores will be collected from each group of respondents according to selected variables summarized to aid data analysis, hypothesis and statistical instrument for the justification of the hypotheses.

A proposed CBSE social science syllabus for 10th level

Unit 2. Contemporary India II (Compulsory Unit) Periods 60

S/N	Themes	Objectives
1.	Resources and Development: Types- natural and human; Need for resource planning, natural resources, land as a resource, soil types and distribution; changing land-use pattern; land degradation and conservation measures.(Chapter 1)	• Understand the value of resources and the need for their judicious utilization and conservation.
2.	Solid Waste Management Skills: Definition, Sources, Waste generators, effects of waste (Health, Economic), and waste reduction techniques. (Chapter 3)	• Understand Solid waste management Skills as well as develop awareness and knowledge to live a clean environment.
3.	Water Resources: Sources, distribution, utilization, multi-purpose projects, water scarcity,	• Understand the importance of water as a resource as well as develop awareness towards

	need for conservation and management, rainwater harvesting. (One case study to be introduced) (Chapter 3)	• its judicious use and conservation.
4.	Agriculture: Types of farming, major crops, cropping pattern, technological and institutional reforms; their impact; contribution of Agriculture to national economy- employment and output. Note: Content of pg no. 44-47 of NCERT Textbook is to be deleted. (Chapter 4)	Understand the importance of agriculture in national economy. • Identify various types of farming and discuss the various farming methods; Describe the spatial distribution of major crops as well as understand the relationship between rainfall re-gimes and cropping pattern. Explain various government policies for institutional as well as technological reforms since independence
5.	Minerals and Energy Resources: Types of minerals, distribution (Note: on map only) use and economic importance of minerals, conservation, types of power resources: conventional and non-conventional, distribution and utilization, and conservation. (Chapter 5)	Discuss various types of minerals as well as their uneven nature of distribution and Explain the need for their judicious utilisation. • Discuss various types of conventional and non-conventional resources and their utilization.
6.	Manufacturing Industries: Types, spatial distribution (Note: on	Discuss the importance of industries in the national economy as well as understand the regional disparities which resulted due to concentration of

	map only) contribution of industries to the national economy, industrial pollution and degradation of environment, measures to control degradation. Note: Content mentioned on page no. 74-75 of NCERT, Geography Text book i.e. Aluminium Smelting, Chemical Industries, Fertilizer Industry, Cement Industry is not required to be deliver in class room during instruction.	industries in some areas. • Discuss the need for a planned industrial development and debate over the role of government towards sustainable development.
7.	Life Lines of National Economy: Importance of means of Communication and transportation, Trade & Tourism (Chapter 7)	To explain the importance of transport and communication in the ever shrinking world. • To understand the role of trade in the economic development of a country.

A proposed Solid Waste Management Skills in the Environmental Science CBSE syllabus for 12th level

Unit 8. Solid Waste (Compulsory Unit) Periods 8

S/N	Themes	Objectives
1.	Solid Waste Management Skills: Components of solid waste management, Management Solid waste management, Purpose of Solid waste management, Solid waste management Skills-: handling, collecting, disposing, storage, transporting, recycling, dumping etc. and employment in solid waste management.	Explain solid waste management: Components of solid waste management, Management Solid waste management, Purpose of Solid waste management, Solid waste management Skills-: handling, collecting, disposing, storage, transporting, recycling, dumping etc. and employment in solid waste management.

REFERENCE

Agarwal, A., Singhmar, A., Kulshrestha, M., & Mittal, A. K. (2005). Municipal solid waste recycling and associated markets in Delhi, India. Resources, Conservation and Recycling, 44(1), 73-90.

Alonso Maria Gaviota Velasco Perez & Themelis Nickolas (2011). Generation and Disposition of Municipal Solid Waste in Mexico and Potential for Improving Waste Management in Toluca Municipality. Waste-to-Energy Research and Technology Council (WTERT).

Asmawati D, Nor B.A.K & Fatimah Y (2012) Environmental Awareness and Education: A Key Approach to Solid Waste Management (SWM) – A Case Study of a University in Malaysia.

Carolina, G. H., & Dewitt, D. B. (2003). Development and security in South East Asia. Aldershot, UK: Ashgate

Desa, A., Kadir, N. B. Y. A., & Yusooff, F. (2012). Environmental Awareness and Education: A Key Approach to Solid Waste Management (SWM)–A Case Study of a University in Malaysia. In Waste Management-An Integrated Vision. InTech.

Dictionary, C. (2008). Cambridge Advanced Learner's Dictionary.

Donkor, F. (1996). An assessment of the learning outcomes of environmental studies in senior secondary schools of cape coast. Unpublished thesis, University of cape coast.

Ekere, W., Mugisha, J., Drake, L., 2009. Factors influencing waste separation and utilization among households in the Lake Victoria crescent, Uganda. Journal of Waste Management 29, 3047–3051

Essays, UK. (November 2013). Formal and Non Formal Education Essay. Retrieved from https://www.google.co.in/?vref=1

Freduah, G. (2004). Problems of solid waste management in Nima, Accra. University of Ghana, Legon.

Gopal G.V & Anand V.V. (2005). Environmental Education in school curriculum an overall perspective, Regional institute of education, Mysore 6.

Gupta, S., Mohan, K., Prasad, R., Gupta, S., & Kansal, A. (1998). Solid waste management in India: options and opportunities. Resources, conservation and recycling, 24(2), 137-154.

Harold k. (1957) the board of Directors and management. Mc Graw.

Hazra T & Goel S. (2008) Solid waste management in Kolkata, India: Practices and challenges, www.elsevier.com/locate/wasman

Hornby, A. S., Wehmeier, S., & Ashby, M. (1995). Oxford advanced learner's dictionary (Vol. 1430). Oxford: Oxford university press.

http://www.businessdictionary.com/definition/solid-waste management.html

https://www.mapsofindia.com/maps/cities/india-map-with-cities.html Retrieved on 25-08-2018

Huysman, M. (1994). Waste picking as a survival strategy for women in Indian cities. Environment and Urbanization, 6(2), 155-174.

Jatput, R., Prasad, G.,& Chopra, A. K. (2009) Scenario of solid waste management in present

Jones, D. (2011). Cambridge English pronouncing dictionary. Cambridge University Press.

Kansal A. (2002). Solid Waste management Strategies for India. Indian Journal of Environment Protection, 22(4): 444-448

Kaushal, R. K., Varghese, G. K., & Chabukdhara, M. (2012). Municipal solid waste management in India-current state and future challenges: a review. International Journal of Engineering Science and Technology, 4(4), 1473-1489.

Kironde, J.M.L., (1999). Dar es Salaam, Tanzania. Onibokun, A.G. (Ed). Managing the Monster. Urban Waste and Governance in Africa. Pp. 101-172. Ottawa, IDRC.

Kothari, D. S. (1966). Report of the Education Commission 1964-66. Ministry of Education. Government of India.

Kumar et.al (2017). Challenges and opportunities associated with waste management in India. Royal Society.

Kumar S, Bhattacharyya JK, Vaidya AN, Chakrabarti T, Devotta S, Akolkar AB., (2009): Assessment of the status of municipal solidwaste management in metro cities, state capitals, class I cities, and class II towns in India: an insight. Waste Management, 29:883–95.

Lakshmi G.V.S, Shubbaiah G.L, Rao D. B. (2006). Methods of teaching Environmental Sience, Discovery Publishing House, New Delhi, India.

Larijani, M., & Yeshodhara, K. (2008). An empirical study of environmental attitude among higher primary school teachers of India and Iran. Journal of Human Ecology, 24(3), 195-200.

Leblang R. (2017).Waste management integrated approach needed for successful diversion and recycling, www.thebalance.com

Levins R, Lewontin R. (1994). Holism and reductionism in ecology. Capitalism, Nature and Socialism 5(4): 33–40

Mayor, M. (2009). Longman dictionary of contemporary English. Pearson Education India.

Miller, D. (1983) A reflection on EO research and some suggestions for the future. Entrepreneurship Theory and Practice.

Ministry of Environment and Forests (MoEF). (2015). The Gazette of India. Municipal solid waste (Management and Handling) rules, New Delhi, India.

Monney I (2014, May 27) Ghana's Solid Waste Management Problems: The Contributing Factors And The Way Forward, Modern Ghana.

Monney I. (2016). Rethinking solid waste management in Ghana, 18-09-2016.

Nachmias, C. F., & Nachmias, D. (1996). Research designs: cross-sectional and quasi-experimental designs. Research methods in the social sciences, 125-151.

Narayana T. (2009). Municipal solid waste management in India: from waste disposal to recovery of resources? Waste Management 929(3), 1163 – 1166.

Patil .C. S & Biradar (2009). Environmental Education, A.P.H. Publishing Corporation, New Delhi, India.

Pervez, A., and A. Kafeel. "Impact of Solid Waste on Health and the Environment." Special Issue of International Journal of Sustainable Development and Green Economics (IJSDGE) 2 (2013): 1-2.

Ramesh C. (2005). Academic Dictionary of Education, Isha books. Delhi india.

Ravindranath, M. J. (2007). Environmental education in teacher education in India: experiences and challenges in the United Nation's Decade of Education for Sustainable Development. Journal of education for teaching, 33(2), 191-206.

Ray, A. (2008). Waste management in developing Asia: can trade and cooperation help? The Journal of Environment & Development, 17(1), 3-25.

Retrieved April 3, 2018, from https://www.merriam-webster .com/dictionary/influence

Shao Alan T. (1999). Marketing Research, South-Western College

Sharma, H. D., & Gupta, A. D. (1995). The objectives of waste management in India: a futures inquiry. Technological Forecasting and Social Change, 48(3), 285-309.

Singh Y. K. (2010). Dictionary of Education, APH Publishing Corporation, New Delhi, India.

Solid waste management BusinessDictionary.com. Retrieved March 12, 2018, from BusinessDicionary.com

Sujauddin, M., Huda, M.S., Rafiqul Hoque, A.T.M. (2008). Household solid waste characteristics and management in Chittagong, Bangladesh. Journal of Waste Management 28, 1688–1695.

Talyan, V., Dahiya, R. P., & Sreekrishnan, T. R. (2008). State of municipal solid waste management in Delhi, the capital of India. Waste Management, 28(7), 1276-1287.

Taylor Frederick W. (1911). Principles of scientific management. The floating Press and its licensors.

Tchobanoglous, G., Theisen, H. and Vigil, S.A. (1993) Integrated Solid Waste Management: Engineering Principle and Management Issue. McGraw Hill Inc., New York.

The Glossary of Education Reform by Great Schools Partnership (2013), https://www.edglossary.org/curriculum/.

Tsiboe, I. A., & Marbell, E. (2004). A look at urban solid waste disposal problem in Accra, Ghana. Unpublished master's thesis submitted to Roskilde University.

Types of Education. (2018). The Iloveindia website. Retrieved 08:31, Mar 14, 2018, from http://lifestyle.iloveindia.com/lounge/. Read more at

UNEP, 200. India: State of the Environment 2001, United Nations Environment Programme, pp.133–149, Chapter 12

Zerbock, O. (2003). Urban solid waste management: Waste reduction in developing nations. Written for the Requirements of CE, 5993.

Zhu, D., Asnani, P. U., Zurbrugg, C., Anapolsky, S., & Mani, S. K. (2007). Improving municipal solid waste management in India: A sourcebook for policymakers and practitioners. The World Bank.

Zurbrugg, C. (2002, November). Urban solid waste management in low-income countries of Asia: How to cope with the garbage crisis. Presented at the Scientific Committee on Problems of the Environment, Urban Solid Waste Management Review Session, Durban, South Africa.

Appendix I

Supporting pictures of places of unmanaged solid waste in India:

1

2.

3.

4.

5.

Source: https://www.youthkiawaaz.com/2017/06/why-is-india-so-dirty/

Appendix II

CBSE SOCIAL SCIENCE SYLLABUS FOR 10TH LEVEL

Unit 2: Contemporary India - II

1. Resources and Development:

Types - natural and human; Need for resource planning, natural resources, land as a resource, soil types and distribution; changing land-use pattern; land degradation and conservation measures. (Chapter 1)

3. Water Resources:

Sources, distribution, utilisation, multi-purpose projects, water scarcity, need for conservation and management, rainwater harvesting. (One case study to be introduced) (Chapter 3)

4. Agriculture:

Types of farming, major crops, cropping pattern, technological and institutional reforms; their impact; contribution of Agriculture to national economy-employment and output. (Chapter 4)

5. Minerals and Energy Resources:

Types of minerals, distribution (Note: on map only) use and economic importance of minerals, conservation, types of power resources: conventional and non-conventional, distribution and utilization, and conservation. (Chapter 5)

6. Manufacturing Industries:

Types, spatial distribution (Note: on map only) contribution of industries to the national economy, industrial pollution and degradation of environment, measures to control degradation.

7. Life Lines of National Economy:

Importance of means of Communication and transportation, Trade & Tourism (Chapter 7)

YOUR KNOWLEDGE HAS VALUE

- We will publish your bachelor's and master's thesis, essays and papers
- Your own eBook and book - sold worldwide in all relevant shops
- Earn money with each sale

Upload your text at www.GRIN.com and publish for free

GRIN